Moray Firth Dolphins

Tim Stenton

Bassman Books

Published by Bassman Books, Burnside Cottage, Newhall, Balblair, Dingwall, IV7 8LT

First published in 2013

A catalogue record for this book is available from the British Library

ISBN 978-0-9567908-3-5

Printed by Big Sky, The Press Building, 305 The Park, Findhorn, Forres, IV36 3TE

Contact the author at timstenton@hotmail.com Visit his website at www.TimtheWhale.com

Also available from Bassman Books

Eilean Dubh – The Black Isle by Andrew Dowsett, James A Moore and Russell Turner
The colour and grandeur of this special part of the Scottish Highlands is brought into focus by this
superb collection of images. 120 pages 140 photographs ISBN 978-0-9567908-0-4

A Cat Called Tess by Russell Turner
Join Tess the cat on her adventures in the woods and at home in her cottage. A book for young
children and cat-lovers of all ages. 40 pages 20 photographs ISBN 978-0-9567908-2-8

Rosemarkie People and Places by Freda Bassindale
This series of photographs, from Victorian times to the Fifties, proves that every village has a
story to tell, whatever its size. 128 pages 160 photographs ISBN 978-0-9567908-4-2

www.russellturner.org

Dedicated to the memory of Beth Dawes 1973-2011

Contents

Foreword 6
1 Introduction 8

2 Dolphins 10
  Cetaceans 10
  Worldwide Distribution . . . 10
  Moray Firth Distribution . . 15
  Size 15
  Morphology
  and Colouration 18
  Feeding 20
  Salmon 21
  Swimming and Diving . . . 27
  Breathing 31
  Breeding 33
  Parentage 34
  '*Rainbow*' and her Calf . . . 40
  Gender 42
  Dolphin Senses 44
  Breaching 46
  Bow riding 52
  Social Activities 56
  Intelligence 61

3 The Moray Firth 62
  Description of Area 63

Where to watch 62
Inner Moray Firth
and Beauly Firth 63
Outer Moray Firth
and Moray Coast 66
Heading South 67
4 Threats 69
  Pollution 69
  Entanglement 71
  Overfishing 71
  Boat Traffic 72

5 Research 73
  Acoustic Location 73
  Fin Identification 73

6 Reporting sightings 78

7 Protecting the Moray Firth
  Dolphins 80
  Special Area
  of Conservation 80
  Dolphin Space
  Programme 80
  Scottish Marine Wildlife
  Watching Code 81

8 Dolphinaria and
swimming with
dolphin programmes 82

9 Other Marine Mammals . . . 84
  Harbour Porpoise 84
  Minke Whale 86
  Short-beaked
  Common Dolphin 88
  Fin Whale 90
  Humpback Whale 90
  Long-finned Pilot Whale . . 90
  Eurasian Otter 91
  Common Seal 92
  Grey Seal 93

10 Photographing Dolphins . 94

11 FAQs 98

12 Sources of Information . 100

Acknowledgements 101

References 102

About the Author 104

Foreword

I first met Tim Stenton many years ago on a whale watching trip in Baja California, Mexico. He was the man resolutely standing outside on deck, searching for whales and dolphins from the moment the sun came up until long after it had set ('you can still see them in the dark', he once pronounced excitedly, 'at least... if there is bioluminescence'). He was never without a camera and a long lens around his neck and never without a broad smile on his face. Most of all, he never took his eyes off the sea, for fear that he might miss something; I'm sure he must have been distracted once or twice, maybe to eat or drink, but I swear I never saw it.

I knew even then that, despite all the attractions of Baja, his heart really lay much closer to home. It was the bottlenose dolphins of the Moray Firth that stirred his emotions and drew him back time and time again. It's easy to forget that we have some wonderful (and surprisingly accessible) cetaceans right on our own doorstep and, as Tim frequently points out, these particular (and exceptionally large) bottlenose dolphins are among our greatest wildlife treasures.

I do understand his passion for the place: I, too, have returned many times since my first visit, to Chanonry Point. I'd been told to park in the spacious car park, stroll down to the water's edge, wait until the rising tide was in mid-flow, and then look out to sea. Scotland's famous bottlenose dolphins would be no more than a stone's throw away. It all sounded too good to be true but, sure enough, there they were just offshore: fishing, performing spectacular aerial displays and doing all the things that dolphins are supposed to do. I've been back countless times since (I have to warn you - it is addictive) and, over the years, have enjoyed some of my all-time favourite dolphin sightings from that unprepossessing little spit of land.

So, needless to say, I was delighted when I heard that Tim was planning to share his infectious enthusiasm in this much-needed book. If, like me, you've been captivated by the Moray Firth dolphins for years (and even if you haven't seen them before - perhaps you weren't even aware they were there?) it will tell you all you need to know about the animals and their home, where to see them and how to support essential efforts to protect them.

Oh, and if you ever spot Tim and his camera with its long lens anywhere on the shores of the Moray Firth - and you happen to catch him taking his eyes off the sea, even for a second, please let me know.

Mark Carwardine
www.markcarwardine.com

1 Introduction

I'm sitting in a hide on a nature reserve somewhere in England. Seeing my camera gear and long lens, people ask if I mainly photograph birds. I explain that my preferred subject is cetaceans, and tell them about my regular visits to the Moray Firth. "Dolphins, dolphins – I didn't know you get them in Scotland," is a common response.

Despite extensive coverage in the national media, and TV in particular, it is a popular misconception that you have to go to Florida or other tropical location to see them. Many people are surprised not just that the Moray Firth has a healthy population but also that it is one of the best places in Europe to regularly see them up close from the shore. That is the inspiration for this book.

Records suggest that the Bottlenose Dolphins have been present in the Moray Firth for about a hundred years although it is only relatively recently that they have attracted large numbers of visitors, making an important contribution to the local economy.

Their continued presence is not guaranteed. Human-induced threats include overfishing, pollution and disturbance through coastal development, boat traffic and exploration for oil. All have a significant impact on the long term viability of the population.

Unlike in many areas, the Moray Firth dolphins can be readily viewed from land – indeed this often offers the best views, although there are also excellent boat-based opportunities. With a little patience, several sites provide a high chance of a sighting, particularly in summer. This is genuine eco tourism with low-impact and low-carbon emissions, and unlike many other forms of wildlife watching the presence of people on land should have no impact on the animals' behaviour.

Waiting at Chanonry Point - watching dolphins is a social activity and can get very busy at times.

The aim of this book is to provide both the casual visitor and the experienced wildlife watcher or photographer with a greater insight into these charismatic creatures, to give an overview of their biology, habitat and behaviours, and to offer information to help ensure their long term survival.

Those who know the area will recognise my bias towards the Inner Firth. This merely reflects the time I spend there. There is something special about having a group of dolphins all to yourself, in stunning scenery. This is much more likely to happen on the Moray or east coast.

Finally, a word on the images used in this book. With a few exceptions, all have been taken by myself during my travels across the world to see cetaceans. All pictures are from the Moray Firth except where stated.

Bottlenose Dolphins breaching close to shore at Chanonry Point, Scotland. In the sun they show a brown colouration though in dimmer light they are grey. This image was taken in the evening with the light behind the camera.

2 Dolphins

'Dolphins' – the very word brings a smile to the face. What do we really know of these engaging, complex animals that share many characteristics, both good and not so good, with ourselves.

Cetaceans

Dolphins are mammals who give birth to live young which they feed on milk. They breathe air but are completely aquatic. Normally the only time they leave the sea is when they breach or jump.

Dolphins belong to the order known as cetaceans, pronounced 'see-tay-shuns'. Comprising dolphins, whales and porpoises, they evolved from land-dwelling mammals some 50 million years ago. The science of cetacean

The excitement of dolphin watching. The dolphin chose to follow the boat which was tacking past Chanonry Point.

classification is constantly developing with some 85-90 species currently recognised.

Cetaceans are split into two sub-orders:

Mysticetes – the baleen whales including the Blue Whale, the largest animal ever to have lived on earth.

Odontocetes – the toothed whales including dolphins and porpoises.

The terms whale, dolphin and porpoise are all used to describe various species of cetaceans. Generally whales are larger than dolphins which are larger than porpoises – but there are many exceptions. Bottlenose Dolphins belong to the family Delphinidae, the oceanic dolphins.

Bottlenose, bottle nosed or bottlenosed? All are in use, highlighting the problems of using common names. The Latin name *Tursiops truncatus* identifies them as Bottlenose or more specifically Common Bottlenose Dolphins. Their closest relatives are the Indo-Pacific Bottlenose Dolphin (*Tursiops aduncus*) and the recently identified Burrunan Dolphin (*Tursiops australis*). The Common Bottlenose is the only one of the three to be found in UK waters. Some scientists recognise some sub-species of *T. truncatus* and hence it is likely that in the future further species of bottlenose dolphin will be identified.

To avoid confusion I have used the term Bottlenose Dolphin to mean the Common Bottlenose Dolphin other than where explicitly stated.

Worldwide Distribution

There are around 40 species of dolphin inhabiting all of the world's oceans and some of the larger rivers. Through dolphinaria and TV programmes such as

Blue whale (*Balaenoptera musculus*) the largest animal ever to have lived. A Bottlenose Dolphin could sit comfortably on each tail fluke. Sea of Cortez, Baja California, Mexico

'Flipper', Bottlenose Dolphins are perhaps the best known species. Bottlenose Dolphins are found in most of the world's temperate and tropical seas and oceans, ranging as far south as the Cape of Good Hope and New Zealand's South Island. Water temperature is probably the primary factor limiting their distribution.

Bottlenose Dolphins can be seen all around the British Isles. Resident populations occur in: The Shannon Estuary, Ireland; Cardigan Bay, Wales; the Inner Hebrides, Scotland; the Sound of Barra in the Western Isles of Scotland; as well as the Moray Firth. In England they are seen mostly on the southern and western coasts from Poole to Cornwall. The Moray Firth population numbers around 200 animals, the population deemed stable or increasing. Although this may seem like a large number of animals, the area which they inhabit is huge.

Common Bottlenose Dolphin in Sea of Cortez, Baja California, Mexico. It looks very different from those found in the Moray Firth but both are the same species.

Common Bottlenose Dolphins, Azores. These are oceanic dolphins living in the open Atlantic.

Common Bottlenose Dolphin in Cardigan Bay, Wales.

Common Bottlenose Dolphins in the Sound of Barra, Outer Hebrides, Scotland. A small population is resident here.

Common Bottlenose Dolphin with a distinctly dark cape or upper back. Baja California, Mexico.

Common Bottlenose Dolphin (*Tursiops truncatus*) Maldives.

Indo-Pacific Bottlenose Dolphin (*Tursiops aduncus*) Maldives.

Common Bottlenose Dolphin, Bay of Biscay. Picture: Hugh Harrop.

Bottlenose Dolphin, Shannon Estuary, Ireland. Home to a resident population similar in size to that of the Moray Firth.

Bottlenose Dolphins off Tynemouth. *Picture: Will Dawes*

Moray Firth Dolphin Distribution

Our understanding of the range of the Moray Firth dolphins is ever increasing. Southwards they are resident around Aberdeen and the Tay Estuary near Dundee. Individuals from the population have been seen as far south as Whitby, North Yorkshire, though they are not resident at such latitudes. Northwards they range up the eastern coast towards Thurso but are rarely seen along the north coast in the Pentland Firth.

Whilst often quoted as being the most northerly population in the world, the Moray Firth dolphins are actually the most northerly in the UK.

Size

The Moray Firth bottlenose dolphins are large. An adult male can grow up to at least 3.9m (13ft) and weigh 650kg. This is more than twice the length and five times the weight of a large adult man. Bottlenose dolphins living in warmer seas, are typically smaller and much lighter (250kg). An increased body mass is useful in conserving heat, very important when the sea temperature rarely rises above about 13C even in summer.

Bottlenose Dolphin with raised tail, Moray Firth.

The Bottlenose Dolphins of the Moray Firth are large, helping them to keep warm in cold seas. They weigh up to 650kg – more than twice the weight of a fully grown tiger.

A breaching Bottlenose Dolphin showing its sizeable girth. Moray Firth, Scotland.

Morphology and Colouration

The shape and colour of Bottlenose Dolphins vary considerably across the globe and even under different lighting. Typically those in the Moray Firth are grey or blue-grey particularly in subdued light. In bright sunlight they can appear brown. All have a pale or white underside. Other populations in warmer waters are often a much lighter grey though they may have a darker cape or upper back.

The Moray Firth dolphins are distinctly chunky, some would say fat, compared to their relatives in warmer waters. They are less streamlined and more barrel like, particularly the slightly larger males. Due to their size, their beaks can also appear proportionately shorter and stubbier.

Above: Bottlenose Dolphin facing away from the camera, lying on its side. Baja California, Mexico.

White underside of a Bottlenose Dolphin in the Moray Firth, Scotland.

The stubby beak of a Bottlenose Dolphin, Sound of Barra, Outer Hebrides, Scotland. This area supports a population of 12-15 individuals living within a small home range between Barra and Eriskay / Uist.

Bottlenose Dolphin feeding on migrating Atlantic Salmon in the Moray Firth.

Feeding

Bottlenose Dolphins feed primarily on fish and cephalopods (including squid), taking a single prey item at a time between their elongated forward pointing jaws. These are equipped with up to 76-98 peg-like teeth designed to grip.

Worldwide, dolphins exploit a wide range of habitats and food sources, displaying a variety of feeding techniques.

The Moray Firth dolphins feed on a variety of fish including mackerel, herring and the Atlantic Salmon which they often pursue in a spectacular fashion. Small fish are swallowed when taken. Larger prey items such as salmon are swallowed head first, enabling the dorsal and other fins of the fish to collapse. This prevents them snagging in the dolphin's throat.

Big salmon take a lot of swallowing. It can take a dolphin 15 minutes or more to swallow a fish, during which time it may be expelled from the mouth and retaken several times. Dolphins often lie on the surface, pushing back their necks and raising their heads when swallowing.

To fuel their active lifestyle in cold waters, bottlenose dolphins eat between 3-6% of their body weight each day.

Atlantic Salmon on migration up the Rogie Falls on the Blackwater River via the Cromarty and Moray Firths.

Salmon

Atlantic Salmon are migratory fish. Born in rivers, they spend most of their lives in the sea before returning, normally to the same river, to spawn. During this migration, often close to shore along historic routes, the salmon are at their most vulnerable. Several rivers drain into the Moray and Beauly Firths, their fish stocks passing through the Chanonry Narrows on migration.

Its topography makes Chanonry Point a perfect place for dolphins to feed.

Due to the topography of the seabed, the returning salmon congregate in the deep water on the western side of the firth, close to Chanonry Point. Combined with the narrowness of the channel at this point it makes it one of the best spots for land-based dolphin-watching in Europe.

Salmon are large fast-swimming fish. Pairs or groups of dolphins may co-operate to catch them – increasing the overall success rate.

Bottlenose Dolphin feeding on Atlantic Salmon in the Moray Firth.

Bottlenose Dolphins feeding on Atlantic Salmon in the Moray Firth. Salmon often leap out of the way or are tossed into the air.

This dolphin expelled the fish before finally swallowing it. Shown in order.

Bottlenose Dolphin chomping on an Atlantic Salmon. Moray Firth, Scotland.

Bottlenose Dolphins put their heads back when swallowing large prey items.

Swimming and Diving

In common with all cetaceans, dolphins' tail flukes are horizontal. They move by contracting and expanding the powerful muscles in their tail stock to produce vertical movement, propulsion being provided only on the upstroke, when the muscle contracts. Bottlenose Dolphins are fast swimmers, reaching speeds of around 25mph. This is only sustainable for short bursts. Their normal speed when travelling or foraging is much slower.

Compared to some other cetaceans such as the Sperm Whale, which can descend to the ocean depths of over two miles, Bottlenose Dolphins are shallower divers. Their maximum depth is around 500m though most dives are much shallower. Typical duration is around two minutes with a maximum of about eight minutes.

Many of the larger whales raise their flukes clear of the water before deep diving, although interestingly only a small proportion of Blue Whales exhibit this behaviour. Bottlenose Dolphins may raise their flukes before diving – more often they do not.

Like all cetaceans, Bottlenose Dolphins swim by moving their tails vertically. Baja California, Mexico.

Bottlenose Dolphin, exhaling on the surface. Sound of Barra, Outer Hebrides, Scotland.

Raised tail flukes of Bottlenose Dolphin. Moray Firth, Scotland.

Bottlenose Dolphins can travel at high speed. Moray Firth, Scotland.

The blow of the Bottlenose Dolphin is sometimes visible. Below: The brief moment when a blowhole is fully open, both Moray Firth.

Breathing

As mammals, living in the sea presents some difficult challenges – not least being the need to breathe on a regular basis. Whilst cetaceans share a common system of breathing with all mammals, they have evolved to suit the demands of their environment.

Over time, the nostrils or blowholes have moved to the top of the head – the uppermost part of the body. This maximises the opportunity for taking a breath. Like all toothed cetaceans, Bottlenose Dolphins have a single blowhole located centrally on the top of their head. Their digestive tract is not linked to their blowhole – they cannot breathe through their mouths.

Like most terrestrial mammals, humans are involuntary or automatic breathers. We breathe when we are

Dolphins are mammals and breathe air like us. Blows are not normally visible except in cold weather or when backlit. Chanonry Point, Scotland.

asleep or even unconscious. For a dolphin who can only breathe when at the surface, this is not an option. They have evolved to become voluntary breathers, requiring conscious brain activity to respire. This makes sleeping difficult. Cetaceans have evolved the ability to close down half of their brain at a time, enabling them to rest, though not in a state that we would describe as sleep. Dolphins cannot afford to lose consciousness – if they do they can't breathe so they would suffocate.

Their blowhole has to be clear of the surface when respiring. This can present problems, particularly in rough seas. Cetaceans complete the respiratory cycle – opening the blowhole; exhaling; inhaling; and closing the blowhole – in as short a time as possible. The blow, when visible, comprises millions of tiny water droplets.

In the short period of the respiratory cycle, up to 90% of the air in the lungs is expelled and replaced. For comparison – in humans it is around 15%.

Dolphins also have the ability to extract a greater proportion of the oxygen in the air than humans, further extending the interval between breaths.

Mother and young calf, Moray Firth.

Breeding

The breeding cycle of the Bottlenose Dolphin is not dissimilar to that of humans. Mothers give birth to a single, large offspring, after a long gestation period, and invest considerable time in raising it.

Dolphins mate belly to belly, with copulation being brief. Gestation lasts approximately 12 months, with most calves in the Moray Firth being born in the summer. Birth occurs in the water – often in a sheltered area. An older female may also be present.

At birth, Bottlenose Dolphin calves are around 1.4m in length, weighing up to 30Kg. They are born tail first with their fins crumpled. Since they must be able to swim immediately, their fins soon harden, allowing them to get to the surface assisted by their mother to take their first breath.

When very young, calves have a floppy appearance – almost as if their skin is too big for their body – this soon disappears.

Whilst in the womb, calves are partially folded. When

Dolphin calf with the foetal folds clearly visible, Moray Firth.

born they have distinct foetal folds, apparent as vertical stripes, along the body. These fade with time. In some populations of Bottlenose Dolphins they are gone within a few weeks – in the Moray Firth dolphins they may still be visible after a year.

Like other marine mammals, dolphin milk is high in fat and proteins. It is squirted into the calf's mouth, so feeding periods can be brief. Calves continue to suckle for about two years, though they are eating solid food well before weaning. They remain with their mothers for much longer – as with humans their degree of independence gradually increasing through adolescence. Some calves stay with their mother even after the birth of a younger sibling. Fathers appear not to take any specific role in parenting though they may remain in the vicinity.

Bottlenose Dolphin calves are lighter in colour than adults, often having distinct yellow patches of skin. This may be due to a variety of factors including jaundice and diatoms – a type of algal phytoplankton. Typically it has disappeared by the time calves are a year old.

Calves face a number of challenges for survival with only around half reaching adulthood. Causes of infant mortality include birth defects, lack of food and disease.

Parentage

Since mothers stay close to their calves, repeated observations of a pair indicate a maternal relationship. Determining paternal parentage is extremely difficult. Since female dolphins are likely to mate with multiple partners, the only reliable method is using DNA testing.

DNA testing has shown that genetically the Moray Firth population is more closely related to those in Cardigan Bay, Wales, than it is to those on the west coast of Scotland.

Dolphin calf with a yellow patch as found on many calves. The exact cause is unknown, Moray Firth.

Bottlenose Dolphin, mother and calf.

Bottlenose Dolphin, calves.

Bottlenose Dolphin, calves.

Bottlenose Dolphin, calves.

Bottlenose Dolphin, calves.

'*Rainbow*' and calf, 15 August 2012.

'*Rainbow*' and her Calf

Every year a number of calves are born into the Moray Firth population. One estimate for the 2012 season is 13 (Source: CRRU). It is particularly special when a well-known individual gives birth.

'*Rainbow*' is a mature female who is frequently seen in the Inner Firth – she is also one of the animals that you can adopt through Whale and Dolphin Conservation's, 'Adopt a Dolphin' scheme. '*Rainbow's*' last calf, named '*Raindrop*' (b. 2005), died in infancy in 2010. It was therefore surprising to see her with a new born calf in August 2012. From the study of photographs taken from Chanonry Point, it was determined that the calf was born between around midday on the 6th and late afternoon on the 7th August 2012.

'*Rainbow*' and calf, 5th September 2012.

'*Rainbow*' and calf, 4th September 2012.

'*Rainbow*' and calf, 14th August 2012.

'*Rainbow*' and calf, 10th August 2012.

'Breeze' clearly a male Bottlenose Dolphin calf! Moray Firth.

Gender

From most angles, determining the gender of a Bottlenose Dolphin is not easy. Adult males are generally bigger than females but this sexual dimorphism is slight. A clear view of the underside and the genital area is required.

Dolphin genitalia are hidden within the body cavity, accessible via the genital slit. Males have two slits. The longer, anterior (front) slit contains the genitals whilst the smaller posterior (rear) slit houses the anus.

Females have a single slit housing genitals and anus, however they also have an additional two slits, (one either side of the main slit) for the nipples.

Male Bottlenose Dolphin showing the two in-line genital slits. Moray Firth.

Female Bottlenose Dolphin, with the single genital / anal slit and the two mammary slits clearly visible. Moray Firth.

Dolphin Senses

The senses of a Bottlenose Dolphin are superbly adapted to their marine environment. Unlike oceanic species they can live in deep or shallow water, offering good or poor visibility.

Sound and Echolocation

From a human perspective at least the most remarkable sensory system of dolphins is that of echolocation.

Sound is crucial in enabling dolphins to understand the environment in which they live. Dolphins can listen to the sounds of the ocean. These can be natural including: movement of tides or the seabed, prey items, potential predators, and other dolphins. Man-made sounds include: watercraft, sonar, explosions, and disturbance of the seabed.

As well as listening, dolphins also produce their own sound in the form of clicks. When the clicks meet a solid object they are reflected back. This system is known as echolocation.

The clicks can be varied in relation to: intensity (loudness), frequency (pitch) and number per second. The time taken for them to return is used to determine the distance to an object, as well as size and density. Dolphins use this information to produce a 'picture' of their environment using sound.

The clicks are produced in the airways in the head – before being magnified and directed by an organ of fatty tissue in the front of the head, known as the melon. The primary receptor for echolocation is a complex mechanism based around the lower jaw which in turn transmits the sound to the inner ear.

Research has found that this is incredibly sophisticated, allowing them to produce a highly detailed map of their surroundings as well as finding prey items at a range

The clicks produced during echolocation emanate from the front of the head or melon.

of over 100m. The echolocation is extremely sensitive, allowing dolphins to differentiate between pairs of identically shaped objects whose size differs by less than 1mm.

Patterns of clicks change depending on the activity. When travelling or foraging they produce a steady series of clicks. These become more frequent and louder when chasing individual prey items.

The frequency range of clicks overlaps with that of human hearing. They can be heard from the bow of a boat and, when dolphins are around in the ocean though not necessarily close enough to see, they can be heard and felt.

Smell

Unlike some fish species such as the larger sharks, which may share their environment, Bottlenose Dolphins have a poor sense of smell.

Touch is an important part of dolphin senses and communication.

Sight

As we know when we try to see underwater, light waves behave differently in water compared to air. Bottlenose Dolphins can see well in both media. Dolphins' eyes are located on the sides of their head just to the rear and above the upper jaw, which provides excellent peripheral vision. The eyes have a large dark brown pupil surrounded by a white sclera, the apparent size of the pupil compared to the sclera varying. They have the ability to adjust the lens cornea to allow for the differing refractive properties of light in air and water.

Like cats, dolphin eyes have a light reflecting layer behind the retina – known as a tapetum lacidum. By allowing the light to reflect through the retina a second time, a dolphin's ability to see in low light conditions is enhanced.

Touch

Touch is very important in Bottlenose Dolphin society. This is apparent particularly when observing them below water, all parts of the body apparently being used.

Taste

Dolphins are thought to have a well-developed sense of taste, being able to differentiate between sour, salty and sweet tastes.

Breaching

Nothing characterises dolphins more than their propensity to leap or breach clear of the water, often achieving a height of several metres – returning to the sea either smoothly, almost without disturbing the surface, or with a huge splash. This behaviour, together with bow riding, spyhopping, and playing with objects both man-made and natural, is perhaps why we find dolphins so endearing.

Why do dolphins breach? The behaviour is not fully understood but it is likely to be for a variety of reasons. It can be a warning signal.

I vividly remember being on a boat on a still evening in the haar or coastal fog. We came across a group of feeding dolphins in the Outer Firth. One of the large males breached repeatedly very close to our small boat as if warning us to keep away.

For oceanic dolphins, breaching increases the distance that they can see. Particularly important when finding food in the open ocean. Breaching is also an important part of social interaction in dolphin society – if two groups of Bottlenose Dolphins meet, breaching often occurs.

Bottlenose Dolphin breaching in the haar. Moray Firth, Scotland.

Breaching Bottlenose Dolphins. Moray Firth, Scotland.

Breaching Bottlenose Dolphins. Moray Firth, Scotland.

Breaches can result in a variety of positions. Moray Firth, Scotland.

Breaching Bottlenose Dolphin, Moray Firth.

Head on view of breaching Bottlenose Dolphin, Moray Firth.

Bow riding

One of the more unusual behaviours shown by dolphins is bow riding. Large boats or ships create a pressure wave as they travel through the water. A dolphin riding in this wave uses less energy than it would if travelling in the open sea. Effectively they get a 'free' ride.

However, dolphins will alter their direction to bow ride a boat only to give up a few minutes later. So why do they do it? Is it because they can?

If the numbers around the vessel exceed the available space, individuals will seemingly jostle for position, seeking the optimum spot. Dolphins also bow ride on the waves created by the larger whales.

Wake riding is similar to bow riding except that it occurs in the wake of a boat. In calm seas with a large group of dolphins it can result in spectacular breaching behaviour.

If you are fortunate enough to be on a boat with dolphins all around you it is important to keep a straight course and maintain a constant speed to reduce the risk of injury to the dolphins – particularly from the propeller (if you have one).

The best to place to be when watching bow riding dolphins – the bow, of course. Baja California, Mexico. I'm at the front by the anchor!

Bottlenose Dolphin bow riding on a tanker. Moray Firth, Scotland.

Bow riding Bottlenose Dolphin in an unusually calm sea. Sound of Barra, Outer Hebrides, Scotland.

Wake riding Bottlenose Dolphins. Sea of Cortez, Baja California, Mexico.

Social Activities

Whilst Bottlenose Dolphins, particularly mature males, can be seen on their own, they are predominantly social animals living in small groups, the typical group size being 2-20 animals, though larger aggregations are not uncommon.

Such groups are not permanent. Individuals come and go, though the nucleus may remain constant. This is known as fission-fusion society. Within the Moray Firth dolphin population some animals seem to spend all of their lives within a certain area, whilst others seen regularly frequenting a location may disappear only to re-appear sometimes many years later.

Bottlenose Dolphins are social animals with highly complex group dynamics, their behaviour being typical of intelligent, social mammals, with large brains.

Within the Moray Firth dolphin population some animals seem to spend a relatively small amount of time feeding, leaving leisure to engage in social activities.

Bottlenose Dolphin breaching. Moray Firth.

Dolphins can sometimes form very close-knit groups. Moray Firth.

Bottlenose Dolphins often play with objects such as this seaweed. Moray Firth.

A dolphin follows a boat, to the delight of those watching.

Play is an important part of dolphin society, individuals often taking an interest in objects, both natural and man-made. When two groups meet, breaching and other social activities frequently occur.

Interactions between dolphins are not always so gentle. Many show teeth or rake marks from encounters with others. This is also likely to be the cause of some of the markings on individuals' dorsal fins.

Interaction between mother and calf. Moray Firth.

Rake or teeth marks from interactions with other Bottlenose Dolphins. Moray Firth.

Bottlenose Dolphin breaching. Moray Firth.

Bottlenose Dolphins socialising. Moray Firth.

Intelligence

One of the most contentious areas of dolphin behaviour is intelligence. How do you measure it? The difficulties are compounded by the fact that dolphin intelligence manifests itself in forms differing from that of humans.

Bottlenose Dolphins have brains (1.6kg) that are slightly larger than those of humans (1.4kg), but even these are much smaller than those of an elephant or Sperm Whale. If the brain-to-body-mass ratio is considered then the positions are reversed. Large brain size is, however, not always an indication of intelligence.

Dolphin intelligence shows itself in several ways. In common with humans and a very few species of great apes, Bottlenose Dolphins have an awareness of self. They recognise themselves as individuals as opposed to being a dolphin.

Imagine showing a dolphin an object that can be seen. It is behind glass so echolocation won't work. The same object is then placed in a container so that it can't be seen but can be scanned using echolocation. Dolphins have the ability to recognise that the objects are the same.

Bottlenose Dolphins show cognitive behaviour – the system of mental processes, comprising: memory, using language, reasoning, problem-solving and decision-making.

The most recent research proposes that dolphins should be considered as non-human persons or 'alien' intelligence.

Bottlenose Dolphins are one of the most intelligent animals on the earth.

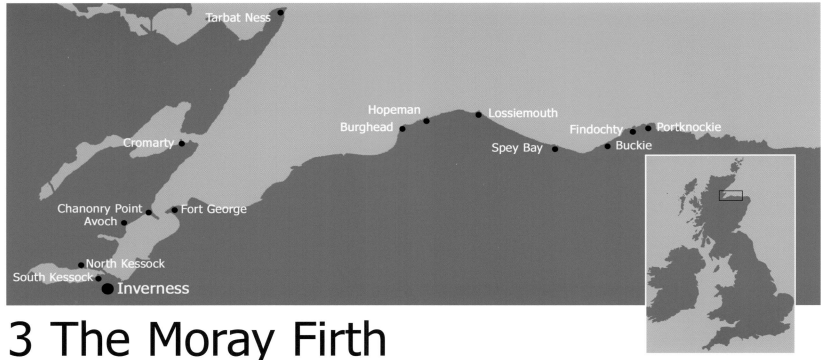

3 The Moray Firth

Where is the Moray Firth?

Imagine a map of Scotland. The large wedge shaped area of sea in the north east includes the Moray Firth. At the Firth's western end lies the city of Inverness. The area from there to Chanonry Point and Fort George is the Inner Firth, whereupon it becomes the Outer Firth which is the open sea. To the south is Aberdeen, whilst the northernmost tip of mainland Britain is in the other direction. Dolphins of course don't recognise such distinctions ranging both south and north, along the east coast of Scotland.

Where to watch

The best place to watch is where the dolphins are! This is everywhere and nowhere. Most areas of the Moray Firth are probably visited by dolphins at some time, however even on a calm day, with good visibility, it is possible to spend a long time watching without a sighting. Dolphins are not strictly territorial although individuals may favour certain areas. Following the habits of the now deceased 'Kess', her daughter 'Kesslet' and grandson 'Charlie' are often to be found in the Inner Firth between Chanonry Point and Inverness, sometimes venturing up the River Ness almost into the city centre.

Any prominent viewpoint along the Moray Firth or Moray Coast will provide a sighting at some time – you may have a long wait, and it might be distant, but it's always worth it. These are some of the best sites:

Bottlenose Dolphin in the Kessock Channel. From North Kessock looking south towards Carnac Point.

Inner Moray Firth and Beauly Firth

Inverness, Carnac Point (off Kessock Road) – on the west side of the mouth of the River Ness, in Inverness. Easy access and can provide very close views of dolphins in the Kessock Channel.

South Kessock – on the south side of the Beauly Firth from Kessock Road. At the end is the now disued ferry slipway.

North Kessock – on the north side of the Beauly Firth provides an easily accessible route, with raised viewing points along Oakleigh and Point Roads. The road ends at the lifeboat station right under the Kessock Bridge. Good facilities with a hotel and shops too.

Avoch (pronounced 'och') – the home of boat operator Dolphin Trips Avoch. Whilst dolphins travel through on a daily basis, land-based viewing opportunities are limited – great for boat trips though.

Bottlenose Dolphin in the mouth of the River Ness. Taken from the marina with Carnac Point in the background.

Chanonry Point from Raddery over a mile and a half away. Judging by the lack of people there, no dolphins were to be seen.

Chanonry Point – widely accepted as one of the best places in Europe to see Bottlenose Dolphins. They can be seen at any time but a rising tide in mid-flow often provides the best opportunities. Despite a large car park, due to its popularity it can become very busy in summer. Many of the images in this book were taken from here.

Chanonry Ness, which ends in Chanonry Point, is a roughly triangular piece of land approximately 1.5 miles long and a mile wide at its base tapering to around 100m. Viewed from the Kessock Bridge, Chanonry Point blocks the direct route to the Outer Moray Firth. At the tip of the point the beach shelves steeply to a depth of around 20 metres. The route past Chanonry appears to be favoured by salmon migrating through the Moray Firth – hence the dolphins, which can often be seen feeding there.

Inner Moray Firth at dawn looking east from Inverness marina. The plume is from the Norboard works near the airport.

Fort George from the Black Isle. Though less than a mile away it takes nearly an hour by road from Chanonry Point.

South Sutors at the entrance to the Cromarty Firth.

Fort George – opposite Chanonry Point, provides views of both the Inner and Outer Firth, with dolphins often close in below the walls. An interesting place to visit even if no dolphins are around.

Cromarty – the front or Shore Street provides a good view towards the Sutors where the Cromarty Firth opens up into the Moray Firth. South Sutors provides a high vantage point with views across to Burghead and beyond but trees partially obscure the view. The access road is single track and badly potholed so recommended for cars only.

The excellent EcoVentures run wildlife watching trips from their base near to the harbour. Dolphins can also sometimes be seen from the Cromarty to Nigg ferry.

Lossiemouth looking east over the River Lossie.

Outer Moray Firth and Moray Coast

The south side of the Moray Firth or the Moray Coast provides numerous viewpoints including:

Burghead – watch from either the old coastguard station on the raised headland or from the shore below. Great for Minke Whales too, but these are normally distant.

Hopeman – the harbour wall can be a good vantage point, but if you have time, a walk east towards Lossiemouth and past the old Covesea coastguard lookout along the scenic coastal path provides greater opportunities for sightings.

Lossiemouth – offers good watching points both ways along the coast but not from the same point. Looking north from the road near the Moray Golf Club Clubhouse is a great vantage point, as is the road overlooking east beach.

Spey Bay – The mouth of the famous Spey River - best known as the home of scotch whisky - is a magical

Burghead offers a raised viewpoint with easy access to the shore.

place. The river enters the sea via an ever changing beach, which is often covered in uprooted trees swept down the river. A great place for seals – otter and osprey are regularly seen too. Don't forget to visit Whale and Dolphin Conservation's award winning Scottish Dolphin Centre.

Buckie – dolphins can often be seen off the harbour in the area of the 'Mucks', a line of three large rocks

Tarbat Ness on the northern shore of the Moray Firth.

The mouth of the River Spey at Spey Bay. A great place to visit even if you don't see any dolphins.

barely visible at high tide lying approximately 500m off shore. Also try the rocks off Craigenroan, below Strathlene Golf Course at Portessie, good for seals too.

Findochty (pronounced 'Finechty') – the best vantage point is from the hill on which sits the prominent white church to the east of the harbour.

Portknockie – from the road above the harbour dolphins can often be seen leaping and passing by. Better still, walk east along the cliff top path to the impressive Bow Fiddle Rock with views over Cullen Bay. In spring and early summer the cliffs are teeming with sea birds. This is also a very good location for seeing minke whales feeding close to land.

Tarbat Ness – On the north side of the Outer Moray Firth one of the best viewpoints is from the lighthouse at Tarbat Ness.

Bow Fiddle rock near Portknockie.

Heading South

Aberdeen Harbour – the south side of Aberdeen harbour along Greyhope Road, Torry offers excellent albeit sometimes distant views of dolphins in the river mouth. Good views from the Shetland ferry too.

Dundee – particularly around Broughty Ferry on the eastern side of the city. Boat trips also run from here.

Bottlenose Dolphin breaching at Spey Bay.

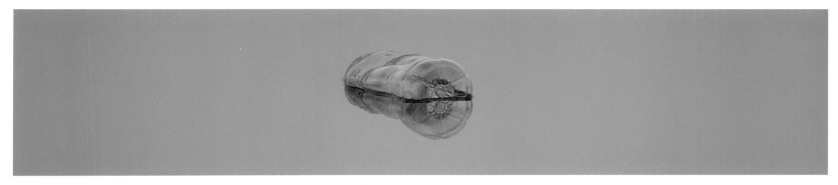

Plastic in our oceans presents an increasing and on-going threat to marine life.

4 Threats

Bottlenose Dolphins have few natural predators. Orca, False Killer Whales and some of the larger shark species will take some, particularly calves. Whilst Orca have never been observed hunting Bottlenose Dolphins in the UK, they have been seen chasing White-beaked Dolphins in the Pentland Firth, Scotland. Bottlenose Dolphins are rarely seen there, unlike Orca – perhaps there is a causal link?

The primary threats to dolphins are man-made, comprising a variety of forms.

Pollution
The effects of pollution are complex but arguably pose the greatest threat to the long-term viability of the population of Bottlenose Dolphins in the Moray Firth. Pollution manifests itself in several ways.

Chemical Pollution
The effects of chemical pollution incidents can be immediate, for example the BP Deepwater Horizon disaster in the Gulf of Mexico – a catastrophe that had a devastating effect on marine life, being responsible for the immediate deaths of many dolphins, as well as legacy effects which are not yet fully understood.

Many chemical substances are discharged into, or end up in, the marine environment. Of particular concern are organochlorides. These include pesticides such as DDT and other industrial compounds such as poly chlorinated biphenols or (PCBs) used in transformers. These compounds accumulate upward throughout the food chain. Cetaceans being at or near the top of the food chain have particularly high levels. Organochlorides being soluble in fat are passed down through the generations – both directly to offspring and to calves via their mother's milk.

Organochlorides are proven to reduce the effectiveness of mammalian immune systems thereby making the animals more susceptible to disease. Although many are now banned they are slow to break down so are therefore still present in our seas.

A huge amount of plastic ends up in our oceans. This is broken down by the action of wind, waves and ultraviolet (UV) light, into ever-smaller particles. Inevitably this will be ingested by marine life including cetaceans.

Pathogens from untreated sewage are one of the causes of skin lesions on Bottlenose Dolphins.

Naval forces recognise that high-powered military sonar causes cetacean fatalities.

Plastic is toxic, causing liver damage and reducing reproductive capacity.

Plastic bags can be mistaken for squid and other prey items, blocking digestive tracts and even causing suffocation.

Biological Pollution

In recent years there have been significant and on-going improvements in the treatment of sewage effluent discharged into the sea. Whilst discharges of raw or untreated sewage have decreased, there are still opportunities for human pathogens to enter the marine environment. At high levels these can cause skin lesions and other diseases for marine mammals.

Acoustic / Sonic Pollution

Acoustic or sonic pollution takes the form of pressure waves. Due to its greater density, sound pressure waves travel around four times faster, and lose less energy in water than in air. There are a variety of sources:

Boat traffic – both pleasure and commercial craft.

Construction – for example the installation of marine wind farms requires piles to be driven deep into the sea bed. This results in pressure levels that have the capacity to injure dolphins.

Seismic surveys – comprising a series of underwater explosions, when prospecting for oil and other natural resources.

Detonation of ordnance – mainly in specific areas for example around the Cape Wrath bombing range. This was the cause of the mass Long-finned Pilot Whale stranding in the Kyle of Durness in 2011.

Military sonar – for example that used in the detection of submarines. The US Navy has admitted that this is killing cetaceans. Similar systems are used by the Royal Navy. Every year seaborne exercises take place in The

Fortunately I was able to rescue my toy dolphin when he became entangled in a discarded fishing net washed up on this Scottish beach – but it could have been a real cetacean.

Minch on the west coast of Scotland, an area known for its diversity of cetacean population.

Cetaceans are particularly susceptible to acoustic or sonic pollution which can cause disorientation as well as physical damage.

Bycatch and Entanglement

Like all cetaceans, dolphins can't swim backwards – this increases their susceptibility to become entangled in commercial fishing gear. If a cetacean gets trapped in a net with its blowhole below the surface it drowns.

Bycatch is the 'accidental' capture of cetaceans in nets when fishing. Mostly their bodies are thrown back into the ocean but some are also taken for human consumption. Cetaceans may also become entangled in discarded fishing nets or gear that is floating in our oceans.

It is estimated that worldwide over 300,000 cetaceans are killed each year in this way. To put this in perspective – for every person in Inverness, five cetaceans die of entanglement each year.

Whilst the level of commercial fishing in the Moray Firth has decreased significantly in recent years it is still undertaken elsewhere. Many of the larger cuts and scars on dolphin dorsal fins were probably caused in this way. If you eat fish, line-caught is preferable.

The problem of bycatch is particularly acute for Harbour Porpoise. This is being addressed by the UK Small Cetacean Bycatch Strategy – the effectiveness of which is unknown.

Overfishing

Dolphins compete with man for many fish species, including herring, mackerel and salmon. Overfishing results in depletion of fish with the consequent effects on the marine ecosystem including dolphins.

Human Disturbance, Boat Traffic and Development

Whilst linked to other factors such as pollution, the direct effects of boat traffic can affect dolphin populations. This can take the form of ship or propeller strikes or general disruption to feeding, resting and breeding activities.

Studies on other Bottlenose Dolphin populations, outwith the UK, indicate that frequent encounters with tourist boats can result in a decrease in the amount of time that dolphins spend resting. In the longer term this can lead to a decline in survival of calves and an increase in the interval between pregnancies. Both of which would result in a decrease in population size. Ironically it is primarily the craft that set out to see the dolphins which appear to be causing the problems, shipping engaged in routine activities apparently having little effect.

In the Moray Firth, commercial dolphin-watching operators comply with a prescribed code. Leisure craft, including dinghies, yachts, motor boats, jet skis and canoes, do not. The problem of water borne traffic is particularly acute in defined feeding areas, for example the Kessock Channel and the Chanonry Narrows (between Chanonry Point and Fort George).

Under the Conservation (Natural Habitats) Regulations 1994 it is an offence to intentionally or recklessly:
- kill, injure or capture whales, dolphins or porpoises;
- disturb or harass them.

Whilst harassment is not defined, if you suspect it is occurring report it to the local police. Photos and video are valuable as evidence.

A jet skier rides far too close to dolphins off Chanonry Point, Moray Firth. This was 'harassment' as the rider intentionally undertook repeated close passes by dolphins which were feeding in a localised area.

5 Research

A surprising number of organisations, from private, governmental and the third sector, undertake research on the coastal and marine environment that comprises the Moray Firth. The areas of study are wide-ranging.

Whilst some research is focused on the wider environment there is a need for the presence of dolphins to be recorded, and even for individual dolphins to be identified. Two of the techniques used are acoustic location and fin identification.

Acoustic Location

C Pods are acoustic monitoring and recording devices that can be anchored to buoys or the sea bed. They record the clicks produced by toothed cetaceans (including Bottlenose Dolphins) at ranges of over 1km. Since C Pods can be left in the water for months at a time they are invaluable in detecting cetacean activity in the locality.

Bottlenose Dolphins can survive without a dorsal fin. This was probably removed by a predator. Sea of Cortez, Baja California Sur, Mexico.

Each dolphin has a unique dorsal fin which can be used to identify it.

Fin Identification

The most commonly seen part of a Bottlenose Dolphin is the dorsal fin. This is visible whenever an animal is at the surface. Its shape and size varies between individuals. Some are sickle-shaped or falcate, others are more triangular.

Over time – due to interaction with other dolphins, scraping rocks, and collisions with boats and fishing gear – dolphins may develop nicks or scratches on their dorsal fin. These are invaluable in identifying specific individuals.

Whilst a few animals are easily recognisable, most can only be identified by those who work with the dolphins on a daily basis. Unlike fingerprints, the markings on dorsal fins may change with time so regular re-identification is required.

As Bottlenose Dolphins age, patches of white develop,

due to the absence of pigment. Particularly prevalent around the edges of the dorsal fin, they can be a useful feature in identifying and ageing individuals.

Unlike other larger cetaceans, such as Humpback and Sperm Whales, the underside or ventral surfaces of the tail flukes are not used for identification purposes in Bottlenose Dolphins. Individual dolphins may however have a distinct pattern of nicks on the trailing edge of their flukes.

The Lighthouse Field Station, a part of Aberdeen University, publishes on its website a catalogue of Moray Firth dolphin fin id images. This is partially funded and populated with data by Whale and Dolphin Conservation (WDC). A separate catalogue is maintained by the charity the Cetacean Research and Rescue Unit (CRRU), this is also available online.

Whilst nicks, notches and other damage may be visible from either side, surface markings are not. For each dolphin a picture of the left and right side of the dorsal fin is needed. Ideally these should be taken from a low level, be square on with the light behind the camera – and they should fill the frame. In practice this proves difficult hence lower grade images are used as well.

No dolphins in UK water are tagged or labelled to assist identification – the work of skilled and dedicated researchers ensures that they do not need to be.

Samples of DNA can be obtained from small skin particles that are continually sloughed off a dolphin as it swims. Cetaceans can also be fitted with satellite tracking devices to monitor their movements across the seas, and measure dive depths and times.

The published fin id catalogues are extensive, having been developed over many years. I have included images of some of the individuals commonly seen in the inner firth, the names and id numbers being in accordance with the University of Aberdeen's catalogue.

The Males

Born in 2001, 'Prism' (#815), son of 'Rainbow', is an active male approaching his prime.

With a large cut at the base of his dorsal fin, 'Mischief' (#23) is one of the most recognisable of all of the Moray Firth dolphins.

'*Sundance*' (#105) b. 1990.

'*Nevis*' (#36) was a large, mature male, first identified when photo id was in its infancy. He was documented as killing at least one Harbour Porpoise. '*Nevis*' died in the late summer of 2010, probably from old age, a year after this photo was taken.

'*Trail Scoop*' or '*Scoopy*' (#748).

'*Denoozydenzy*' (#573).

The Females and Calves

'*Zephyr*' (#866) and '*Breeze*' (b. Aug 2009).

'*Moonlight*' (#580) and '*Lunar*' (b. 2010).

'*Jigsaw*' (#30), an old female, shows extensive white fringing around her dorsal fin.

A member of the most famous Moray Firth dolphin family, '*Kesslet*' (#433), b. 1994, (above) is the daughter of '*Kess*' who died in 1998. Her son '*Charlie*' was born in September 2007.

'*Rainbow*' (#31), b. late 1980s, and her new calf, born in August 2012.

'*Charlie*' as a young calf less than a year old.

6 Reporting Sightings

Although there is a healthy population of Bottlenose Dolphins in the Moray Firth they are spread over a wide area. Information on sightings is useful for both dolphin-watchers and scientific purposes. Whilst you don't need to be an expert to participate, it is useful to record and report sightings.

What to report
Location
Date and time
Species – reporting as an unidentified dolphin species is fine if you are unsure.
Number of animals
Additional data including: sea state, visibility, direction of travel and any indications of behaviour are also useful. Photos are invaluable if you have them.

Where to report
Sea Watch Foundation
seawatchfoundation.org.uk – all sightings.
Hebridean Whale and Dolphin Trust
whaledolphintrust.co.uk – sightings on the west coast of Scotland and the Hebrides.

Sources of Sighting Information
Unlike the networks that have developed to record and disseminate bird sightings, those for cetaceans are less developed. Sea Watch Foundation provides the most comprehensive geographical coverage, via its website. Unfortunately this is only updated on a sporadic basis. More up-to-date information is available via social media networks, including for the Moray Firth:
Adopt a Dolphin on Facebook – from Charlie Phillips,

Breaching Bottlenose Dolphin, Moray Firth.

WDC (Whale and Dolphin Conservation).
WDC Adopt a Dolphin on Twitter – from Charlie Phillips, WDC.
Dolphin Centre WDC on Twitter – The WDC Dolphin Centre at Spey Bay.
Cetacean Research and Rescue Unit (CRRU) – on Facebook.

Locally, perhaps the best of way of obtaining up-to-date information is from other dolphin-watchers.

Bottlenose Dolphins, Kessock Channel, Inverness.

7 Protecting the Moray Firth Dolphins

On board the Gemini Explorer, operating out of Buckie.

The Moray Firth dolphins and their habitat are governed by a variety of legislation and guidelines – some of which is complex and open to interpretation.

Special Area of Conservation

The Inner Moray Firth is designated as a Special Protection Area for wildlife conservation purposes.

Due to the presence of Bottlenose Dolphins the Moray Firth contains a Special Area of Conservation (SAC) designated under the EU Habitats Directive, comprising one of the largest Marine Protection Areas in Europe. The SAC protects the inner waters of the Moray Firth, from a line between Lossiemouth (on the south coast) and Helmsdale (on the north coast) westwards.

An objective is that any new activities taking place in the range of this Bottlenose Dolphin population should not affect the integrity of the SAC. This means in particular that the population growth rate should not be affected so as to threaten population viability.

Dolphin Space Programme

The Dolphin Space Programme (DSP) is an accreditation scheme for wildlife tour boat operators. Its aim is to encourage people who go out to observe dolphins and other marine wildlife to "watch how they watch", and to respect the animals' need for space.

All operators in the Moray Firth who run dolphin watching trips are members of and comply with the requirements of the programme. These provide strict guidelines in relation to:
- approaching dolphins particularly in enclosed areas,
- specifying maximum times for encounters, and

• prohibiting the swimming with, and feeding of dolphins.

Adoption of such a code provides protection for the dolphins whilst providing excellent opportunities for boat-based watching. The DSP also has objectives of conservation of marine life through co-operation and economical and ecologically sustainable tourism.

If you go out on a dolphin watch boat trip the objectives of the DSP should be explained.

The Scottish Marine Wildlife Watching Code

Developed by Scottish Natural Heritage (SNH), the Scottish Marine Wildlife Watching Code aims to minimise the disturbance to cetaceans and to make sure that watching wildlife is a good experience for both wildlife and people.

General Protection applicable to all Cetaceans in UK waters

Whilst the DSP only applies to commercial boats, all users of watercraft have a legal duty not to harass dolphins (see page 72). If you suspect dolphins are being harassed then report it to the local police.

Dolphin bow-riding on the 'Saorsa' operated by EcoVentures, Cromarty. As often happens the dolphin approached the boat which was travelling on a straight course at a constant speed.

Responsible dolphin-watching from Dolphin Trips Avoch.

Why go and see captive dolphins when you can reliably see them from land in the UK?

8 Dolphinaria and swimming with dolphins programmes

In recent years lists of '100 things to do before you die' have become very popular. Many of them list swimming with dolphins, often making the 'Top 10'. Most fail to mention the impact on the dolphins...

My first experience of seeing dolphins and orca, or killer whale, was at Windsor Safari Park in the mid '70s – it seemed quite acceptable at the time.

The effects on the health and welfare of animals kept in such facilities are now better understood but perhaps not so widely known to the millions of visitors who flock to them each year.

Why is this? Many dolphin pools seem to be large enough. They contain water that appears clean and blue. Food is a plentiful. Vets are available to look after their health. Surely if the dolphins weren't happy they wouldn't perform and do tricks?

Dolphins are gregarious animals living in large or small groups with complex social interactions. Some individu-

Even the largest dolphinaria pools are a fraction of the size of the open ocean.

als are more dominant than others. In an enclosed area such as a pool there is insufficient room for such interactions to occur.

Dolphins travel over vast distances in search of food. A pool may be large in relation to the animals' size, but even the largest are a fraction of the area of their natural range. They also lack the features, such as rocks and currents, that typify the natural environment.

The skin of a dolphin is surprisingly sensitive. They have evolved to live in salt or freshwater environments. Dolphinaria pools are normally chemically treated with chlorine which causes skin lesions and ulcers.

Captive dolphins frequently suffer from stress, have a shorter life expectancy and produce less-healthy offspring that those living in the wild. Even dolphins kept in sea pens are not in a natural environment – they are often kept in shallow water away from areas which they would normally inhabit.

Where do captive dolphins come from? Some are born in captivity but many others are captured from the wild, transported considerable distances and forced to live in unfamiliar environments. Dolphin hunts are incredibly cruel. In Taiji, Japan over 2,000 dolphins are brutally killed every year – only a few 'lucky' animals survive – all to provide so-called 'entertainment'.

What about swimming with dolphins in open water? Whilst much less harmful than programmes involving captive dolphins, the ethics of swimming with wild dolphins is contested. Opponents cite the transfer of disease, and the fact that human disturbance may affect dolphin behaviour.

It is illegal to approach, with the intention of swimming with, any cetacean in UK waters.

Typical view of a distant Harbour Porpoise. They are normally only seen on the calmest of days. Chanonry Point, Moray Firth.

9 Other Marine Mammals

The Moray Firth is a rich and diverse ecosystem supporting a number of mammals that are either marine, or in the case of the otter, found along the coast. Some such as the mighty Fin Whale are rare visitors, but others such as the Minke can with some patience and favourable weather conditions be seen.

Harbour Porpoise (*Phocoena phocoena*)
The diminutive Harbour Porpoise is the most common cetacean both in UK waters and the Moray Firth. Often mistaken for the much larger Bottlenose Dolphin, they are frequently seen, particularly on calm days, travelling or feeding in small family groups.

Harbour Porpoise, also known as the Common Porpoise or just Porpoise, are small with a length of up to 1.9m and weighing up to 75kg. They have a blunt head and a small triangular dorsal fin with a slightly rounded top. They surface as if on a wheel, showing a distinctive rolling motion. Other than on the calmest days only the dorsal fin is visible. They exhale with a distinctive sound giving rise to their local name of puffing pig. Unlike dolphins they rarely approach boats or breach.

Porpoise primarily feed on small fish, often on the seabed. This makes them particularly susceptible to being caught in fishing nets.

Harbour Porpoise off The Sutors, Cromarty in the Moray Firth.

Harbour Porpoise, off Gairloch on the West Coast of Scotland.

Harbour Porpoise, off Gairloch on the West Coast of Scotland.

Harbour Porpoise in the Moray Firth.

Minke Whale feeding, about 1¼ miles off the Sutors, near Cromarty on the Moray Firth. Flocks of feeding birds may indicate their presence.

Minke Whale
(*Balaenoptera acutorostrata*)

A close relative of the largest animal ever to have lived – the Blue Whale – the Minke or Northern Minke is the smallest of the baleen whales. Growing up to around 10m, it is significantly larger than a Bottlenose Dolphin and much heavier, weighing up to 9 tonnes.

Minke Whales are regularly sighted in the Outer Firth, normally some distance from shore.

Other than size, the surfacing sequence is different to that of a Bottlenose Dolphin with the long back, showing forward of the relatively small dorsal fin. The blow is often not visible.

In the Moray Firth Minke Whales are normally seen in the summer and autumn.

Minke Whale in the Cromarty Firth. The buildings in the background are the fabrication yard at Nigg.

Minke Whale on the West Coast of Scotland, near Gairloch.

Short-beaked Common Dolphin showing the characteristic falcate fin and hourglass pattern of markings on its flanks. Off Gairloch on the West Coast of Scotland.

Short-beaked Common Dolphin
(*Delphinus delphis*)

Belonging to a group known as the oceanic dolphins, the Short-beaked Common Dolphin, having a maximum length of around 2.7m, is smaller and much lighter than the more robust Bottlenose.

Travelling quickly, either singly or in large herds, they are more likely to be seen in the Outer Firth or North Sea, but have been sighted in the enclosed Beauly Firth.

Short-beaked Common Dolphins are often to be found in conjunction with groups of feeding birds, particularly gannets, causing surface disturbances that can be seen from many miles away. Many texts describe a yellow side panel as a diagnostic feature, but the characteristic hourglass pattern on the flanks, and a falcate (sickle shaped) dorsal fin are more reliable means of identification.

Short-beaked Common Dolphins making a rare appearance in the Inner Moray Firth. A small group were regularly seen in the area in the summer of 2006.

Short-beaked Common Dolphin breaching whilst travelling at speed. Off Gairloch on the West Coast of Scotland.

Fin Whale (*Balaenoptera physalus*)

The second-largest animal ever to have lived, the Fin Whale is a very rare visitor to the Moray Firth. Measuring up to 27m, and with mature females weighing up to 120 tonnes, they dwarf the minke whale. They produce a vertical blow up to 7m in height.

Fin Whale — a very rare visitor to the Moray Firth. This was the first ever sighting in the Inner Firth, seen off Chanonry Point in 2007. (Picture: Will Dawes)

Humpback Whale (*Megaptera novaeangliae*)

A rare but increasingly frequent visitor to our shores, Humpbacks are slow moving robust whales. Growing up to 18m, they have extended pectoral fins or flippers. A mainstay of whale-watching in many areas, they frequently breach.

Humpback Whale. Baja California Sur, Mexico. Sightings in UK waters are becoming increasingly common.

Long-finned Pilot Whale Pico, Azores.

Long-finned Pilot Whale (*Globicephala melas*)

Normally seen in family groups, Long-finned Pilot Whales are highly gregarious. They are around 6m in length and are black or dark brown in colour. Long-finned Pilot Whales are seemingly less common than in the past when they were regularly seen in the firth.

Eurasian Otter (*Lutra lutra*)

Normally found close to, and on the shore, the Eurasian Otter prefers shallow coastal areas, particularly kelp beds. Unlike seals and cetaceans it requires access to a supply of freshwater for drinking and grooming.

Truly marine otters which do not need to spend time on land are only found in the Pacific. European Otters, whilst they can be found in the sea, are not the same species as the Sea Otter (*Enhydra lutris*).

Many of the rivers and freshwater lochs around the Moray Firth are populated by otters that may never visit the sea.

Common Seals, Findhorn, Moray Firth.

Common Seal or Harbour Seal (*Phoca vitulina*)

The smaller of the two seal species to be found in UK waters, the Common Seal prefers more sheltered waters than the larger Grey. Colonies can be found at several locations along the Moray Coast, particularly where there are sandbanks.

Common Seals are often described as having a dog-like face with an upturned nose, the nostrils forming a V-shape. Their colouration is highly variable from brown to grey, often with a mottled or pebbled patternation.

Common Seal, Findhorn, Moray Firth.

Common Seal, Findhorn, Moray Firth.

This Common Seal pup came ashore at a busy Chanonry Point – it was tired wanting a rest but was otherwise OK.

Grey Seals, Scotland.

Grey Seal (*Halichoerus grypus*)

Widespread around our coasts, the Grey Seal is the larger of the two resident seal species, males growing up to 2.5m and weighing 150kg. Grey seals have longer faces than their smaller Common Seal cousins – the larger males in particular having a distinctive Roman nose.

The dolphin doesn't have to fill the frame. Particularly in good light, wider or environmental images can be very effective.

10 Photographing Dolphins

With the advent of digital photography when most people have a camera of some sort to hand it's natural to use these to try and capture pictures of dolphins. Getting a photo of a dolphin is normally not difficult – getting a good photo is a lot harder. I'm full of admiration for anyone who can get a sharp dolphin picture with their mobile phone.

A few tips

Dolphins can move very quickly and unpredictably. Expect to miss a lot of shots especially when they are active. You will get plenty of pictures that are out of focus, blurred, or where the animals are only partially in frame or even missing altogether!

Unless the weather is bad, it's always worth setting up

With the camera at ground level the perspective is dramatically altered and the dolphins appear bigger, although it can be difficult to keep the subject in the frame.

your gear as soon as you arrive. Even on a calm day with visibility of several miles, they can seemingly appear without warning.

On a good day you can expect to take many hundreds of pictures of which maybe 90% will subsequently be deleted.

Whilst mobile phones and compact digital cameras can produce excellent results they are frequently hampered by a lack of control, and a lag between pressing the shutter and the shot being taken.

The preferred camera choice is a digital SLR.

- Set to maximum resolution to allow for enlargement (L fine in JPEG).
- Shooting in RAW is recommended if you are familiar with its use and have the time to process images post shooting.
- Set drive to continuous shooting at the highest frame rate possible.
- If using auto focus systems select the continuous auto focus setting for moving subjects or use the sports mode.
- Keep the ISO as low as possible (maybe 400) but shutter speeds need to be high when the animals are active. It also helps to have a smaller aperture (high f number) to increase the depth of field thereby making focusing less critical. My optimum settings are 1/1600 second or higher at F8, ISO 400 when using a 500mm lens.

Dawn often provides some superb light – it is very early in midsummer though.

- Lenses – assuming your digital SLR has a 1.5-1.6x or similar crop factor, compared to 35mm film (i.e. most Canon, Nikon, Pentax and Sony models) the best lens is probably a 70-300mm or similar. This works well both from land and boats. Longer lenses can be used successfully and are needed when the animals are further out, but since they have a narrower field of view you can miss a lot of photos. Fast, i.e. f2.8-f4, 300-500mm lenses are ideal but are very expensive and heavy. Often the action is happening so quickly that you don't have time to change lenses. If you do, be very careful – sand and salt spray will damage equipment.
- If dolphins are close, zoom out to the wider end so that you don't miss any action. You can always crop later on a computer.
- Camera supports are useful when using longer lenses for extended periods. They provide a steady platform and perhaps more importantly take the weight off your arms, as well as allowing you to remain focused on a specific patch of water. I use both a tripod and a monopod. Whilst a tripod can be used in quiet areas a monopod provides greater flexibility and can be used in crowded locations such as Chanonry Point. There is no

need for a complex head on a monopod as the angle of view can be altered by adjusting the height and by tipping it backwards and forwards.
- Spare batteries and cards – if you've got them, take them! So many times I've seen people go back to their cars only to miss the best of the action. If the dolphins are active you can take a huge number of photos. Finally, if animals are around keep on shooting – don't stop to 'chimp' or review your pictures. Keep shooting.
- Make sure you have a waterproof bag or case to protect your gear. A clear plastic bag is useful for protecting your camera when shooting in the rain and is essential when going out on a boat. A lens cloth is useful too.

Light
The best photos are normally taken on sunny days. This gives colour to the sea and the dolphins, as well as providing ample light. The Moray Firth runs approximately east to west, with the views from Chanonry Point – at least in dolphin sighting terms – being generally NE to E. In the morning you tend to be shooting into the sun but by the afternoon it has moved around into a more favourable position.

More abstract images can work well – this was unintentional.

Whilst not ideal conditions it is possible to get atmospheric pictures in thick fog or haar.

Clothing
Even in summer it is usually cold. The number of days I can shoot without wearing a jacket is limited – the need for hats and gloves is frequent.

My Gear
I'm a lifelong Canon user. The images in this book were taken using a variety of bodies including: 1DMk4, 1DMk3, 50D, 40D, 20D and 350D. Lenses used include: 18-200, 17-40 F4L, 70-200 F2.8L IS, 300 F2.8L, 500 F4L IS and 800 F5.6L IS. 1.4x and 2x extenders or teleconverters were used as required. A few of the wide shots were taken with a Canon G9.

A recommendation – if you are investing in an SLR system buy either Canon or Nikon. Unfortunately both are Japanese – I would love to buy a camera system made by a company based in a country that does not hunt whales and dolphins.

Camera complete with 'Nevis' my toy dolphin.

11 Frequently Asked Questions

Q Do dolphins smile?
Bottlenose Dolphins have a fixed facial expression which appears to us to be a smile. They express their emotions in other ways including: tail slapping, showing their teeth, breaching and physical contact as well as a range of vocalisations.

Q Do dolphins mind the rain?
Being marine and hence wet the answer is probably not. Weather conditions may affect their behaviour for example by influencing the location and availability of fish. Observational evidence suggests that dolphins can be very active on the surface on windy days in rough seas. This can be partially explained by their need to be clear of the water surface when breathing.

Q Do dolphins ever come onto land?
Unlike seals and sea lions, dolphins spend their entire lives, including giving birth, in the water. A very small number of Bottlenose Dolphins on the eastern seaboard of the USA intentionally partially beach themselves when catching small fish. This does not occur in the UK. Bottlenose Dolphins can be seen in very shallow water particularly if chasing prey.

Q Do dolphins kill sharks?
The larger dolphin species, particularly Orca or Killer Whales, do kill sharks including the Great White. Dolphins, particularly calves, are also killed by sharks.

Q Are dolphins hunted?
Unfortunately yes. Dolphins are actively hunted in many areas including: Japan, South East Asia, Africa and

Photographing dolphins in the driving rain.

South America. Unbelievably they are still brutally hunted in the Faroe Islands, a part of Denmark and a member of the European Union, only a few hundred miles north of Scotland. In the UK all cetaceans are protected by law.

Q Are dolphins related to humans?
Whilst both humans and dolphins are mammals, they are not closely related. Dolphins' and whales' closest relatives are: pigs, hippos, camels, giraffes, deer and cattle.

Q Do dolphins kill porpoises?
Both Bottlenose Dolphins and Harbour Porpoises are present in the Moray Firth – porpoises seemingly avoiding the dolphins. Whilst Bottlenose Dolphins have been

Eye contact with a wild Bottlenose Dolphin. Sea of Cortez, Baja California, Mexico.

observed killing porpoises, it is not known if this is common or widespread.

Q Do dolphins put on a display only when people are watching?

No – dolphins breach when they want to, the presence, or absence, of watchers apparently not affecting their behaviour. If you are lucky enough to be on a boat, in calm seas, and have Bottlenose Dolphins bow-riding it is possible to make eye contact. I've also tried it with Common Dolphins but they don't seem to engage in the same way.

Q How long do dolphins live for?

Captive dolphins have much shorter life spans than those in the wild. Bottlenose Dolphins in their natural environment live for a maximum of 40-60 years.

Q What is the collective noun for a number of dolphins?

Various terms are used including: pod, pack, school, group and herd. My preference is to use group for smaller numbers and herd for larger aggregations. Fish congregate in schools and Orca in pods.

Q Do dolphins rescue swimmers?

There are a number of recorded instances where dolphins have rescued swimmers, though none in the UK. One theory is that since dolphins are aware of the need to consciously reach the surface to breathe, their natural reaction is to help swimmers in distress.

Q Will I see Kate Humble if I go to the Moray Firth?

Probably not... but you never know.

Kate Humble enjoying the dolphins at Chanonry Point, Moray Firth.

12 Sources of Information

There are numerous organisations involved in the study, protection and conservation of the Moray Firth dolphins and the marine environment:

Whale and Dolphin Conservation (WDC)
www.wdcs.org
WDC run the Seal and Dolphin Centre at North Kessock near Inverness and the Wildlife Centre at Spey Bay. Check out: www.adoptadolphin.com/blog/ the blog of Charlie Phillips their local field officer where you can 'Adopt a Moray Firth Dolphin'.

Sea Watch Foundation
www.seawatchfoundation.org.uk
An excellent source of sightings information, and the organisation to which sightings should be reported.

Moray Firth Partnership (MFP)
www.morayfirth-partnership.org
A voluntary coalition of organisations and individuals the MFP is a forum to share information, plan and help to implement integrated ways of managing the valuable assets of the Moray Firth. The MFP also provides grants for community projects.

British Divers Marine Life Rescue (BDMLR)
www.bdmlr.org.uk
British Divers Marine Life Rescue is an organisation dedicated to the rescue and wellbeing of all marine animals in distress around the UK. They have trained marine mammal medics located across the UK. Rescue Hotline: 01825 765546 during office hrs (07787 433412 out of office hrs).

Hebridean Whale and Dolphin Trust
www.whaledolphintrust.co.uk
Their focus is on the west coast and Hebrides. Any sightings in this area should be reported to them.

Scottish Marine Animal Stranding Scheme
www.strandings.org
The Scottish Rural College (SRUC) runs a marine animal stranding scheme. This project provides a systematic and coordinated approach to the surveillance of cetacean, basking shark and turtle strandings in the UK and to the investigation of causes of death.

University of Aberdeen – Lighthouse Field Station
www.abdn.ac.uk/lighthouse
Undertake research projects on the marine environment. Their fin id catalogue can be found at www.abdn.ac.uk/lighthouse/gallery/

Sea Mammal Research Unit – St Andrews University
www.smru.st-andrews.ac.uk

Scottish Natural Heritage (SNH)
www.snh.go.uk
Scottish Natural Heritage is the government's adviser on all aspects of nature and landscape across Scotland.

Cetacean Research and Rescue Unit (CRRU)
www.crru.org.uk
Based in Gardenstown on the Moray Coast, the CRRU is concerned with research, marine mammal rescue and education. They publish a fin id catalogue online.

Dolphin Space Programme
www.dolphinspace.org
The Dolphin Space Programme (DSP) is an accreditation scheme for wildlife tour boat operators. They provide a list of accredited boat operators.

Police
Inverness Police Station 0845 600 5703
Buckie Police Station 0845 600 5700

Acknowledgements

I would like to thank the following for their assistance (knowingly and otherwise), hospitality, friendship and support during the writing of this book.

Neil MacGregor; Karen van der Zijden; Catherine Clark; Charlie Phillips; Will and Izzy Dawes; James and Gill Moore; Alan Airey; Jane Simons; Kenny and Pauline Majury; Barbara Cheney; Russell Turner; Mark Carwardine; Rachel Ashton; Hugh Harrop; Steve, Di and Annabel Williamson; and of course my lovely Emma …… not forgetting all of my other friends and family with whom I have watched cetaceans across the globe.

References

Whales, Dolphins and Porpoises Naturally Scottish – Gillham, K. and Baxter J. publ. Scottish Natural Heritage (2011)

Whales, Dolphins and Seals – Hadoram Shirihai and Brett Jarrett (2006)

Distribution, abundance and population structure of bottlenose dolphins in Scottish Waters – SNH Commissioned Report No 354 (2011)

Integrating multiple data sources to assess the distribution and abundance of Bottlenose dolphins Tursiops truncatus in Scottish waters – Cheney, B. et. al. Mammal Society (2012)

Coast to coast: First evidence for translocational movements by Scottish Bottlenose dolphins (UK) – Robinson, K. et al. (2009)

Dolphins – Ben Wilson (1998)

Bottlenose Dolphins –Paul Thompson and Ben Wilson (2001)

Collins Wild Guide Whales and Dolphins – Mark Carwardine (2006)

Britain's Sea Mammals – Jon Dunn, Robert Still and Hugh Harrop (2012)

Yorkshire's Whales, Dolphins and Porpoises –Robin Petch and Kris Simpson (2011)

The development of a framework to understand dolphin behaviour and from there predict the population consequences of disturbances for the Moray Firth bottlenose dolphin population – SNH Commissioned Report 468: (2012)

Site Condition Monitoring of bottlenose dolphins within the Moray Firth Special Area of Conservation: 2008-2010 – SNH Commissioned Report 512

Valuing the benefits of designating a network of Scottish MPA'S in territorial and offshore waters – A report to Scottish Environment Link (2012)

Get to know the bottlenose dolphins of northeast Scotland – WDCS Leaflet (Date unknown)

Dolphins – Jonathan Bird (2007)

WWT Species Action Plan – Marine and Freshwater Cetaceans 2012-2020

In Defense of Dolphins – Thomas I. White (2007)

Poster showing all of the world's cetacean species to scale. Available direct from www.ukogorter.com

Flickr.com a photo sharing website has a huge range of images of the Moray Firth dolphins

There are also numerous pages on Facebook

About the Author

'Tim' the person.

A resident of that Scottish county known as Englandshire, Tim Stenton is a long-time and frequent visitor to Scotland, with a particular affinity for the Highlands and Islands, in recent years spending several weeks each summer on the Moray Firth with the specific intention of watching and photographing Bottlenose Dolphins.

Although a graduate in biology he has never utilised this skill in a professional capacity.

A non-award winning photographer, Tim can be readily spotted by the toy dolphin resting on the end of his lens. This, together with an extensive collection of other toy animals, accompany him on his travels across the globe to photograph cetaceans.

Tim's proudest achievement is to have a Sperm Whale living in the Mediterranean Sea named after him.

www.TimtheWhale.com
timstenton@hotmail.com

'Tim' the Sperm Whale in the Ligurian Sea, Italy.